I Spot an INVENTION!

Sani and Jai's Treasure Hunt

Written by
DINESH N. MELWANI

Illustrated by
RANA ALY

GLOBAL BOOKSHELVES
INTERNATIONAL, LLC

Author: Dinesh Melwani
Illustrator: Rana Aly
Editor: Danielle Davis
Editorial Director: Janan Sarwar

Published by Global Bookshelves International, LLC
Louisville, KY 40241
www.GlobalBookshelves.com

To comment on this book by email, send your message to the publisher at globalbookshelves@gmail.com.

Notice
The author has made every effort to ensure the historical accuracy of the inventors and their creations. However, invention origin stories often contain elements of mythology, and the details presented reflect the most accurate publicly available information. Neither the author, any organizations he is affiliated with, nor the publisher can guarantee the ongoing accuracy of this information or be held responsible for any unintentional errors or how the information is applied.

First Edition
Printed in the United States of America
ISBN: 978-1-957242-30-9
Library of Congress Control Number: 2025921593

Dedication

For Jaian and Saanvi:

It's a great big world out there, and it needs your ideas.
Keep questioning, learning, and discovering.

Love,
Dad

On the first day of summer break, Sani and Jai looked forward to their annual outdoor treasure hunt with Dad.

Sani ran over to her brother's bed. "Jai," she shouted, "it's time to w a k e u p!"

Normally a sleepyhead, Jai hopped out of bed and went with Sani to see what Dad had in store.

"Well, I have bad news. It's raining."

"NOOo," Sani moaned.

Jai chimed in, "Could we do a different kind of treasure hunt instead?"

"Great idea!" said Dad.

"An invention hunt!" said Sani.

INVENTIONS are all around you, wherever you are.

These are new things that have been created from someone's imagination—they solve problems, help with chores, and usually make life better.

At the first stop on their invention hunt,
the kids turned on the light.
Dad said, "I spot an invention!"

"Is it the light bulb?" Jai asked.

"Yes! The light bulb is one of the most famous inventions.
Can you imagine life without it?"

"A single idea," said Dad,
"can light up the entire world."

Throughout the 1800s, many people worked on the concept that would become the light bulb.

But it wasn't until 1878 that Thomas Edison created a version that worked well enough for people to use in their stores and homes. Edison was granted a PATENT for his light bulb, and things have never been the same.

When an invention is new, the government can issue a patent to legally protect it from being copied without the inventor's permission.

Sani and Jai continued exploring.

When they got to the bathroom, Dad repeated,
"I spot an invention."

"Bandages? Really?" Sani and Jai asked.

"Yes! A single idea can help heal."

In 1921, Earle Dickson's wife kept cutting her fingers while cooking. She'd tape clunky cloth to her wounds, but it wouldn't stick.

So Dickson experimented with cleaner, more effective bandage materials and designs. By combining a tiny bit of cotton with a sticky, germ-free fabric, he discovered an *INNOVATION* for his wife—and all of us.

His company, Johnson & Johnson, loved the idea and put it into production. They've since sold over 100 billion BAND-AID® brand adhesive bandages.

They continued their treasure hunt in the kitchen.

"I spot *a lot* of inventions!" said Dad.
"Which do you think is most interesting?"

Sani, who liked to help with dinner, picked one right away:
"The roti maker!"

Roti Maker

"That's right, Sani. A single idea can
make it easier to connect around the
table with people you love."

FLOUR

Roti is a round flatbread commonly served with meals in South Asian, Caribbean, and South African cultures.

Pranoti Nagarkar was an **ENGINEER** who struggled to balance her job with her desire to make home-cooked food for her family. There were machines for making coffee and baking bread, but nothing for rotis, which are eaten by over 3 billion people around the globe.

In 2017, after 8 years of designing, Nagarkar was granted the patent for her **ROTIMATIC®** machine. Since its launch, the invention has cooked up over 200 million rotis for families worldwide.

"I spot an invention—something to help clean up this mess we made."

Jai knew that one: "The dishwasher!"

"A single idea can make chores a breeze for hard-working parents—and their kids," said Dad.

With a twinkle in her eye, Sani yelled, "To give them more time to play!"

In the late 1800s, Josephine Cochrane decided she'd had enough of washing dirty dishes by hand. She liked to host parties, which made a lot of them!

So Cochrane, who had **INVENTORS** in her family, got to work in her backyard shed.

She created a machine that used a flat, spinning wheel to spray dishes with soapy water. After a while, they came out sparkling clean. The **dishwasher** was born, a valuable part of many kitchens today.

Indian Recipes

A+ Name Jai
$a^2 + b^2 =$

Sani

After cleaning up, Sani and Jai went to their bedrooms to see what other treasures might be hidden in plain sight.

Dad stood in the doorway:
"I spot an invention."

Sani held up the thing that most annoyed her in the morning: "The alarm clock!"

"Even if you don't like it, a single idea can ensure we're awake to the beauty of every day."

Before **ALARM CLOCKS**, people had very few ways to help them wake up when they needed to. The first simple versions, created in the late 1800s, were too expensive for ordinary people. So George Kern went straight to his clock factory in Illinois to tinker.

By 1913, Kern had a reliable device that was simple and affordable enough for lots of people to keep near their pillows.

Just then, Dad's phone rang.
Mom needed to be picked up from work.

Jai smirked: "I spot another invention!"
He pointed at his father's phone.

"Aha! That's one of the most
world-changing inventions
of all time," Dad said.

"A single idea
can allow us
to talk with
our friends
and families
across town
or even
continents."

Alexander Graham Bell was a scientist and inventor fascinated by sound, hearing, and speech. Both his mother and wife were Deaf, which greatly influenced his life's work.

After others had tried to create a voice-transmitting machine for decades, Bell finally cracked the code in the 1870s.

Though the first PH☎NE was very different from what's in our pockets today, it quickly and vastly changed how people could communicate.

As Dad drove toward Mom's hospital, rain fell like small pebbles on the car. Sani noticed the *swish-swish* wiping it away. "I spot one more invention. Windshield wipers!"

"I didn't even think about that, Sani!" Dad said. "A single idea can help us see clearly when the road ahead is hard to find."

On a snowy New York day in the early 1900s, Mary Anderson had an idea while touring the city in a trolley. Drivers had to stick their heads out their windows to see the road in the snow. What a nuisance!

She devised a lever to control a rubber blade that could clear snow or rain from a vehicle's windshield.

In 1903, Anderson was awarded a patent, and by the 1920s, companies were installing **WINDSHIELD WIPERS** on every new car.

When Mom joined them, Sani and Jai couldn't hold in all they'd learned.

"So what's your favorite thing you've discovered?" Mom asked.

Jai responded first. "Inventions are cool!"

Sani took more time to think before saying,
"A single idea can change the world. And it can come from *anyone*."

Robert Patch was just 6 years old when he was granted a patent for his unique TOY TRUCK design.

Charles Greeley Abbot was 101 when he received his final patent in the field of SOLAR ENERGY.

As soon as they arrived home,
Sani and Jai rushed in!

"I wonder what they're doing,"
Mom muttered.

The kids went straight to the junk drawer and craft box.

"What are you up to?" asked Dad.

Sani looked at her parents as if the answer was obvious:

"We're inventing something!"

ACTIVITIES FOR EDUCATORS AND CAREGIVERS

I SPOT AN INVENTION!

How to Play:

- Look around your environment and say, "I spot an invention that helps us... [listen to music, move from place to place, communicate with friends, etc.]"
- Children guess that invention (e.g., a radio, a bicycle, a phone).
- Encourage discussion about how it was invented and how it has changed over time.

DESIGN YOUR OWN INVENTION

How to Play:

- Give kids a blank sheet of paper and colored pencils, markers, or crayons.
- Ask: "What problem would you like to solve?"
- Children brainstorm an invention that solves a problem, then give it a name and draw it.
- Encourage them to share how it works: "What does it do? How does it help people?"

BUILD IT!

How to Play:

- Give kids cups, straws, paper, tape, and other simple materials.
- Challenge them to build a bridge, tower, or moving object using those supplies.
- Encourage teamwork and problem-solving, just like real engineers!

For more information about this book and additional resources, please visit globalbookshelves.com.

GLOSSARY

- **Invention** – A new thing created by someone to solve a problem or make life easier.
- **Patent** – A special right given to an inventor by the government to ensure no one else copies their idea.
- **Innovation** – A fresh way of doing something or improving an existing invention.
- **Engineer** – A person who uses science and math to design and build things.

Dinesh N. Melwani is a partner in the Washington DC law firm of Bookoff McAndrews and a former professor of patent law. For nearly 20 years, he has represented both US and foreign clients in all aspects of patent protection and counseling. More information about his practice may be found at: www.bomcip.com/our-team/dinesh-melwani/

Rana Aly is a children's book illustrator and character designer based in New York. As a mother, she's passionate about creating diverse characters and inclusive worlds where all children can feel seen and represented.
Follow her journey at ranalystudio.com

www.ingramcontent.com/pod-product-compliance
Lightning Source LLC
Chambersburg PA
CBHW061147030426
42335CB00002B/140